Viviane Caires Pereira
Alex F. Santos

The Expansion of Oil Palm Production in the State of Pará

Viviane Caires Pereira
Alex F. Santos

The Expansion of Oil Palm Production in the State of Pará

Case Studies

ScienciaScripts

Imprint
Any brand names and product names mentioned in this book are subject to trademark, brand or patent protection and are trademarks or registered trademarks of their respective holders. The use of brand names, product names, common names, trade names, product descriptions etc. even without a particular marking in this work is in no way to be construed to mean that such names may be regarded as unrestricted in respect of trademark and brand protection legislation and could thus be used by anyone.

Cover image: www.ingimage.com

This book is a translation from the original published under ISBN 978-613-9-69206-4.

Publisher:
Sciencia Scripts
is a trademark of
Dodo Books Indian Ocean Ltd. and OmniScriptum S.R.L publishing group

120 High Road, East Finchley, London, N2 9ED, United Kingdom
Str. Armeneasca 28/1, office 1, Chisinau MD-2012, Republic of Moldova, Europe
Printed at: see last page
ISBN: 978-620-8-12525-7

Summary

This paper will present some of the impacts related to the expansion of oil palm in the state of Pará, and their consequences, which initially affect small farmers and the entire production chain. Possible mitigating actions will also be presented, which would benefit not only family farmers but also the population in general, agro-industries and the Brazilian economy, seeking to achieve development with a focus on sustainability and social equality.

Keywords: Oil palm, Palm oil, Agroecology.

SUMMARY

CHAPTER 1

INTRODUCTION

In view of all the data collected, it can be seen that the expansion of oil palm was already predicted a few decades ago. The biggest investments in scientific research into oil palm were made by Embrapa. The aspects mentioned here cover part of the problems encountered in the northern region of Brazil, in the state of Pará, knowing that many of the problems of this crop are interconnected with others and are like a cycle of actions and causes. The search for techniques and crops that bring economic growth has, in a way, created an imbalance in the process of family farming.

In this research paper, some of the impacts related to the expansion of oil palm in the state of Pará, their consequences and possible mitigating actions that will benefit not only family farmers but also agro-industries and the Brazilian economy will be presented.

Research procedure

Initially, we delimited the theme and the study area, and from there we began to search for general information on the state of Pará, biotic and abiotic characteristics, socio-economic data from the IBGE, and the main product, Dendê, was researched from its origin to cultivation techniques, pests and biological requirements for its production, as well as the impacts caused by its expansion.

With the information recorded, it was possible to analyse the conditions found, looking for options for improvement.

Objectives

Characterise the current oil palm situation in the state of Pará;

Present the advantages and disadvantages of oil palm expansion; Present actions that can reduce the negative impacts and highlight the positive impacts.

Hypothesis

The problems can be socio-economic, such as the lack of planning and execution for crops in the region. In addition, the laws are poorly considered, disfavouring small farmers and prioritising large agro-industries.

Justification

This research was designed to establish improvements in all the main aspects related to oil palm production in the Amazon region and to conduct economic and social processes in a way that is more comprehensive and less harmful to the environment and human beings.

History of the state of Pará

The military impulse that led the Portuguese to Pará in 1616 was part of an ambitious political project to conquer the Amazon valley, coupled with a no less ambitious economic project to exploit the local biodiversity - the so-called "drugs of the hinterland", Western spices with a high value on the European market.

The foundation of Belém on 16 January 1616 was the first step in this project of territorial conquest. The process was continuous and tense and the occupation of the territory was based on the massacre or enslavement of the indigenous populations and on warlike confrontation with the other European powers that had factories in the Amazon. Little by little, through military expeditions, new regions were annexed: the Guamá, Acará and Mojú river valleys, the lower Tocantins, the Caetés coast (today's Salgado coast), the region of the "Maranhão road", which today we call Bragantina, the Ilha Grande de Joannes, today's Marajó, the Gurupá peninsula, the North Cape, today's Amapá, the lower Amazon, the Xingu and Tapajós valleys, the upper Amazon and the Rio Negro valley.

Each of these territories corresponded to new routes for the exploitation of wealth, whether it was the drugs of the sertão, land to be added to the patrimony of the main settlers, or the enslaved indigenous populations.

This economic pattern, based on extractive gathering, river navigation and the enslavement of Indians, together with the geographical aspects of the equatorial rainforest, ended up differentiating the new colony from the rest of Portuguese America. As a result, from 1626 onwards, there were two Portuguese colonies in America: Brazil, which included the whole of the north-east and the entire southern part of the colony, and Grão-Pará e Maranhão, which included the Amazon, Maranhão and, along with it, Piauí and part of Ceará. What differentiated these two colonies was basically the mode of production: in Brazil, monoculture predominated and, in Grão-Pará, extractive activity.

The Catholic Church was also part of this colonising enterprise. Its missionary work, however, was often in competition with the settlers' mercantile interests, given that the religious missions were often large production spaces, with interests in land ownership, the backlands drug trade and the maintenance of indigenous populations in these production spaces. The debate between the settlers and the church lasted throughout the 17th and 18th centuries, until the new political mentality in force in the metropolis began a process of expropriation of their property and also of dynamising the local economy.

It was the policy of the Marquis of Pombal, the emblematic figure of enlightened despotism in Portugal. Carvalho e Mello, the famous marquis, minister of King José I, came to rule the nation with iron hands after the great tragedy that was the destruction of Lisbon by an earthquake, followed by a tsunami and an immense fire in 1775. His political project favoured a new pattern of mercantilist economy, the slave trade as an instrument to increase production and the valorisation of Portuguese colonies that had previously been decentralised, such as Grão-Pará and Maranhão.

The Marquis of Pombal's interest in Grão-Pará was almost personal: he appointed his own brother as governor of the province, a nephew of his as bishop of Pará and various other relatives and friends to prominent positions in the colonial bureaucracy that concerned this American colony.

He also encouraged the beautification of Belém: the construction of the Governors' Palace (now the Pará State Historical Museum), the rebuilding of the city's main churches - such as the Sé, Santo Alexandre, Carmo, SanfAnna and São Joãozinho - and other public and residential buildings. The Bolognese architect Antônio José Landi stood out in this activity and, like many other specialists brought by Pombal, ended up settling in the colony.

However, the Marquis' dismissal launched Grão-Pará and Maranhão into a cycle of decadence that lasted for almost a century, until around 1870. During this period, the colony experienced many changes: the troubled accession to Brazilian independence, the emergence of the press and an active intellectual life and, above all, the experience of civil war, which was Cabanagem.

The Cabanagem, this popular revolution so poorly understood by history, was a moment of social explosion with a demographic and cultural impact that marked Pará forever. Some historians estimate that a third of the population died in the conflict, which was full of tragic episodes. The movement was led by the lower classes. Beginning in 1835, it took Belém and spread throughout the Amazon. A people's government was installed and remained in force until 1838, when the capital was once again conquered by the legalist forces. However, the conflicts lasted until around 1840.

From then on, Pará began to be heavily penalised by the Brazilian government. A popular, Indianised, subversive, independentist and upstart colony, in the eyes of many, it gradually lost its status. The general language, a mixture of several indigenous languages intertwined with Latin, commonly spoken by all the traders and caboclo populations of the Amazon valley and practically the language of commerce at Ver-o-Peso and the language of politics in the working class neighbourhoods of Belém, was banned. The caboclo populations were often persecuted and the large agricultural or extractive properties no longer received the attention and support of the central government.

Pará grew rich quickly, and the previous essentially diversified economic base quickly gave way to "monoculture" latex extraction. The proclamation of the Republic in 1889 had no major impact on the formation of the local ruling classes - as was the case throughout Brazil, and the same instruments of social reproduction and control were maintained in essence, even with the end of slavery.

With the decline of the two rubber cycles came an agonising stagnation, from which Pará only emerged in the 1960s with the development of agricultural activities in the south of the state. From the 1960s onwards, but especially in the 1970s, growth accelerated with the exploitation of minerals (mainly in the southeast of the state), such as iron in the Serra dos Carajás and gold in Serra Pelada.

Regional survey

The state of Pará is located in the north of the country and is bathed to the north-east by the Atlantic Ocean, where it shares borders with other states: to the south with Mato Grosso, to the west with Amazonas, to the north with Amapá, to the east with Maranhão and Tocantins and to the north-west with Roraima, Guyana and Suriname (Cerqueira et al).

Its territorial extension is 1,247,950.003 km^2 , and it is considered the second largest Brazilian state in terms of area, behind Amazonas. Its area corresponds to approximately 14.6% of Brazil's territory. The state's terrain has low altitudes, with more than 85% of the territory not exceeding 300 metres in altitude. The northern portion is marked by the Amazon plain and the other regions are characterised by depressions and small plateaus, as well as the coastal plain. The highest point is the Serra do Acari, at 906 metres above sea level (Cerqueira et al).

Crossed to the north by the equator, the state has a hot and humid equatorial climate. The average annual temperature varies between 24 °C and 26 °C and the rainfall is approximately 2,800 mm per year. The Amazon rainforest is the predominant biome, but there are also areas of savannah, mangroves and grasslands on the island of Marajó. The hydrographic network, which is very rich, is made up of rivers that form part of the Amazon, Tocantins-Araguaia and North-East Western Atlantic basins. Among the main rivers that flow through Pará are the Amazon, Jari, Pará, Tapajós, Tocantins, Trombetas and Xingu (Cerqueira et al).

Study area

Location

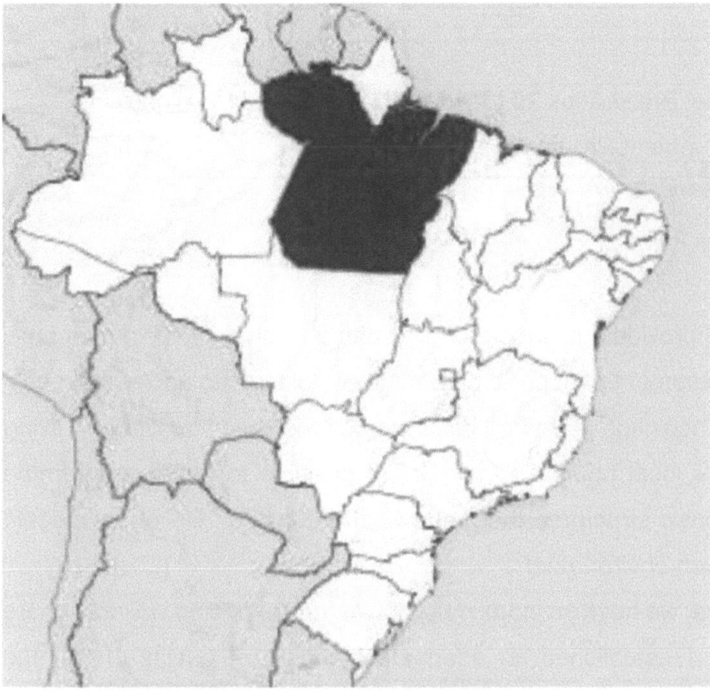

Figure 1 Map 1 - Location of Brazil and the State of Pará

Demographic Data of the State of Pará	
Capital	Bethlehem
Estimated population 2014	8.073.924
Population 2010	7.581.051
Area (km)²	1.247.954,32
Population density (inhab./km)²	6,07
Nominal monthly household income per capita of the resident population 2014 (Reais) (1)	631
Number of municipalities	144

Figure 2 Table 1 - Demographic data for the state of Pará-IBGE 2014

Municipal Human Development Index -	MHDI
MHDI 1991	0,413
MHDI 2000	0,518

MHDI 2010	0,64 6
Source: Brazil Atlas 2013 United Nations Development Programme	nidas

Figure 3 Table 2 - Municipal Human Development Index - MHDI

The MHDI provides guidance for the state on issues of socio-economic growth and development of specific communities, taking into account aspects of profitability, housing, existing basic structures and qualitative living conditions for the local population, making it possible to better programme investments in basic structures that improve the quality of life of the people allocated to a given environment.

In this case, we have an index much lower than expected in Pará, where the proximity of 1 establishes an adequate standard of quality of life and development, but we observe an index much lower than the proportionality of its population growth, even though there has been growth since 1991 until the last survey carried out, as reported in the table above by Atlas Brasil 2013.

CHAPTER 2

SOCIO-ECONOMIC SITUATION OF THE LOCAL COMMUNITY AND BRAZIL

With regard to the socio-economic situation of the traditional community, the vast majority of family farmers earn less than 1/2 the minimum monthly wage per capita from subsistence cultivation (manioc, maize, small animal husbandry, etc.) Palheta et al 2000 confirmed this with IBGE data. The IBGE data showed that the per capita income of urban dwellers was just over one minimum wage in 2010.

Resident population - total	7.581.051	People
Resident population - household situation - urban	68,5	%
Resident population - household situation - rural	31,5	%

Figure 4 Table 3 - Population sample - IBGE 2010

This table shows the total number of people and their percentage of the population in each scenario, highlighting the higher population in the urban area.

Families living in private households and number of family members	1.943.035	Families
Families living in private households and number of family members - Urban	1.399.694	Families
Families living in private households and number of family members - Rural	543.341	Families
Resident population - total	7.581.051	Persons

Figure 5 Table 4 - Household Sample - IBGE 2010

Completing the information in the previous table, now with the number of families, the largest number of families prevails in the urban area.

Per capita income

Average monthly nominal income per capita of permanent private households - totals	R$548,17
Average monthly nominal income per capita of permanent private households - urban	R$664,31
Average nominal monthly income per capita of permanent private households - rural	R$263

Figure 6 Table 5 - Average per capita income - IBGE 2010

Considering that the monthly minimum wage in 2010, obtained from the Brazilian portal, was R$510.00, we can see the precariousness in rural areas.

Indigenous Population in the North of Brazil

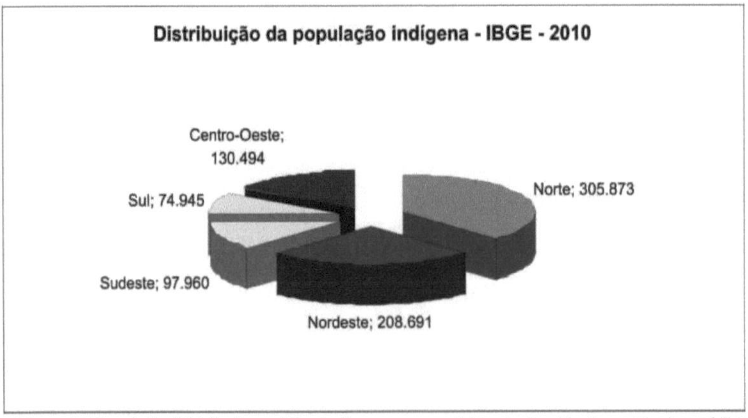

Figure 7 Graph 1 - Distribution of the indigenous population in the Northern Region of Brazil - Funai/IBGE 2010.

The indigenous peoples are the original inhabitants of Pará. Even today, they form societies that help us understand the Amazonian universe, as they are specific peoples with a rich and diverse ethnic and cultural heritage. Current records show the presence of 32 peoples in the territory of Pará, with approximately 16,000 Indians. Most of these groups speak languages from three distinct trunks: Macro-Jê, Tupi and Karib. There are 39 officially recognised indigenous lands, representing 24.52% of the state's total area.

Indigenous Peoples are present in all five regions of Brazil, with the

North region concentrating the largest number of individuals, 305,873 thousand, approximately 37.4 per cent of the total. In the North, the state with the largest number of indigenous people is Amazonas, representing 55 per cent of the region's total (FUNAI 2010).

Quilombola communities

There are already 240 quilombola communities known to exist in the state of Pará. It is believed that many more will be identified .
At first glance, the existence of such a significant number of communities descended from quilombos in Pará may seem strange, given the widespread idea that slavery was not so important in the Amazon. Although the use of black labour in the Amazon did not reach the same figures as in other regions of the country, it was of great importance to the local economy. In the various regions of the current state of Pará, black slaves were used as labourers in agricultural and extractive activities, in domestic work and in urban construction. The history of slavery in Pará was marked by the resistance of blacks and Indians who sought their freedom by escaping, building quilombos and participating in Cabanagem. In the 21st century, the descendants of the quilombos continue their constant struggle for their rights. The main target now is the titling of their hard-won lands. Quilombola men and women in Pará can be proud of their many achievements. It was in Pará, in the municipality of Oriximiná, that a quilombola community received collective title to their land for the first time in 1995. And it is in this state that the largest number of quilombola land titles are concentrated. Since 1998, Pará has had legislation regulating the process of titling this category of land. This innovative legislation guarantees the right to self-identification of communities without the need for an anthropological report - something that the federal government only recognised in 2003. As well as telling a little about this history, from the period of slavery to the most recent conquests, the site also describes the current way of life of some communities. Through these accounts, we are trying to make the daily life, work, festivities,

leisure and struggle of the quilombola women and men of Pará better known[1]
.

Protected areas of the Legal Amazon

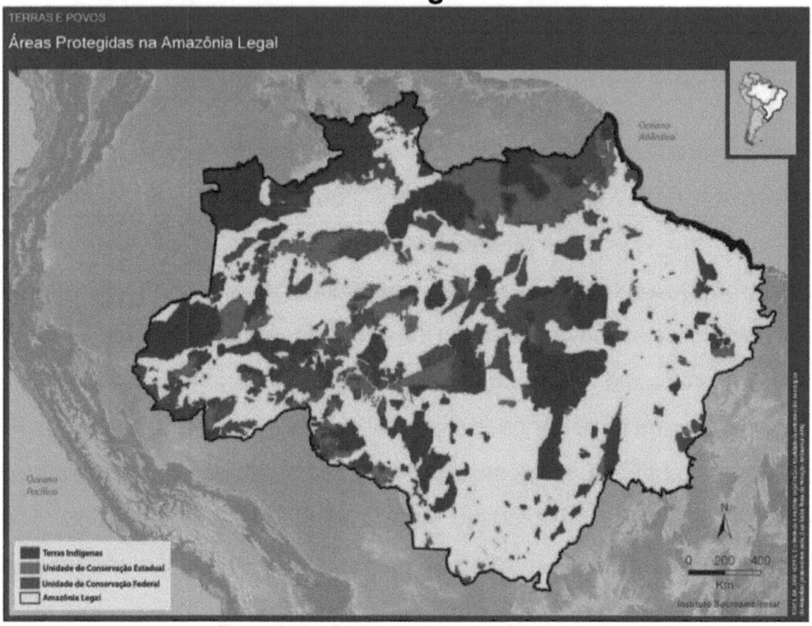

Figure 8 Map 2 - Protected areas of the Legal Amazon

Here we will have a representation of the Legal Amazon, observing indigenous aspects as the main factors.

This image shows the territorial demarcations of protected areas and indigenous territory. Within the Legal Amazon area are state and federal conservation units. The information highlighted gives a positive indication of indigenous use, as their activities are neither predatory nor massive enough to cause major problems for the environment, such as agriculture, livestock and other production activities, especially those that do not take into account soil recovery and reforestation.

[1] Source: http://www.cpisp.org.br/comunidades/html/i_brasil_pa.html

IT classifications according to degree of reforestation pressure

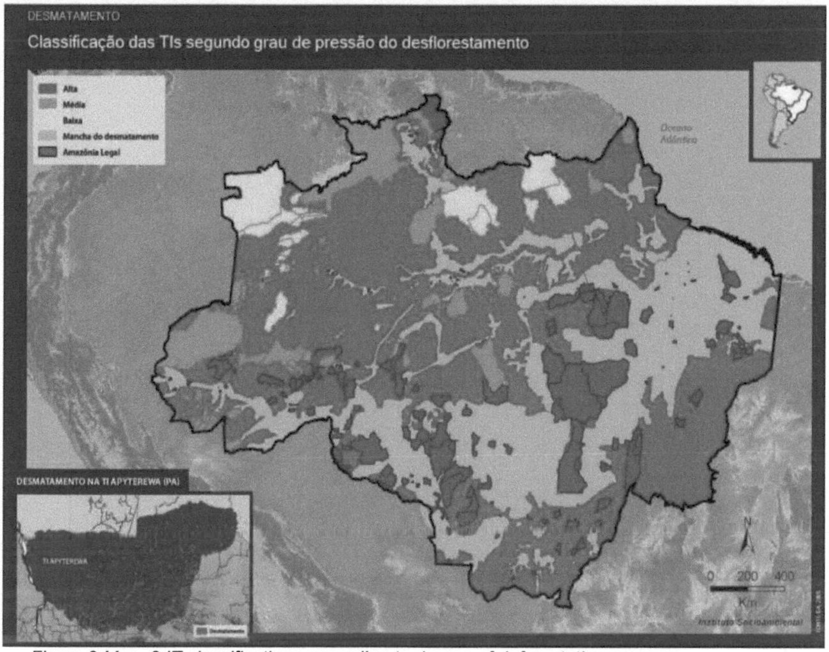

Figure 9 Map -3 IT classifications according to degree of deforestation pressure.

In this image, deforestation in TI APYTEREWA (PA) is more pronounced in areas where occupation is close to the main areas of deforestation that have already taken place. There is no concrete correlation as to how this deforestation occurred, but it points to the possibility that other higher intensity agricultural activities or fires generated them.

Soil in deforested areas in the Legal Amazon

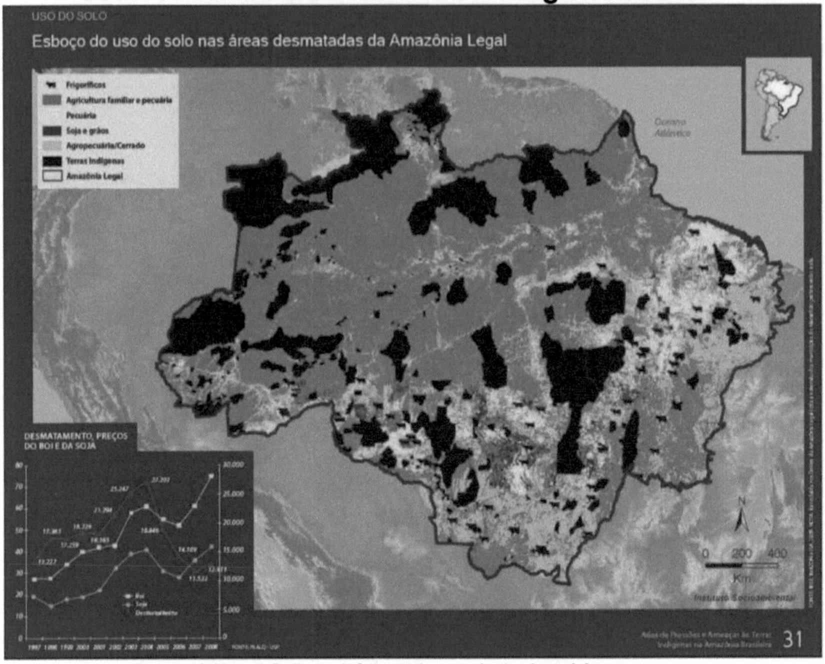

Figure 10 Map 4 Soils in deforested areas in the Legal Amazon

Another important piece of information concerns the main means of production in the Legal Amazon and their potential degradation. The main activities are related to agriculture and livestock farming. Both occupy large areas around the Indigenous Lands and have a high risk of harming the villages, given that there is large-scale deforestation and the use of pesticides that can develop serious diseases and contaminate the soil extensively.

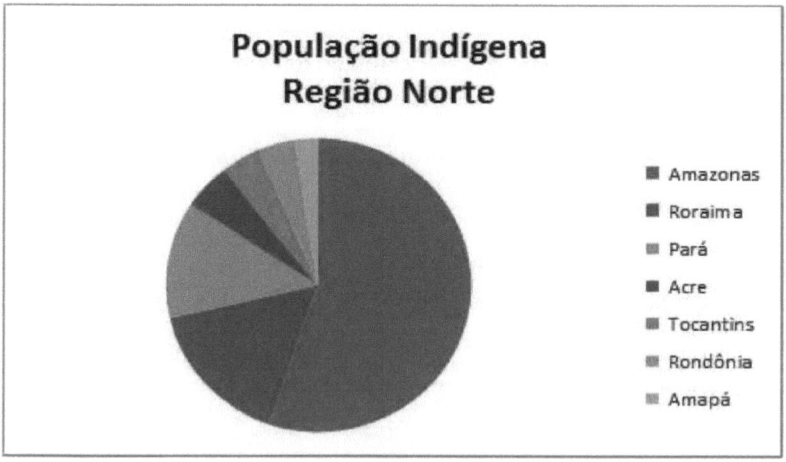

Figure 11 Graph 2 - Comparison of the Indigenous Population in the Northern Region - FUNAI / IBGE 2010.

This comparison shows that the state of Pará is the third state with the largest Indigenous population. (FUNAI 2010), behind the states of Amazonas and Roraima in the northern region of Brazil.

Indigenous peoples are part of Brazil's history and much more than that, they are also part of the earth. Even without specific studies, indigenous people already treated nature with respect.

Indigenous Population IBGE data

Indigenous people	51.217	Person s
People living on indigenous lands	38.036	Person s
Indigenous people living on indigenous lands	35.816	Person s

Figure 12 Table 6 - Indigenous population in the state of Pará - IBGE 2010

This table shows that there has been a migration of indigenous people to other regions, with a difference of 15,401 people from the indigenous population living on indigenous lands, which is less than the indigenous population to be considered in this study. The numbers of non-indigenous people living on indigenous land are in stark contrast to the massive and predatory occupation of areas demarcated for this population. If we take into account that this area is occupied by more non-indigenous people than indigenous people, then we will have a very high cultural and diversified impact, since the direct contact of white people with indigenous people brings frequent damage in terms of possible extinctions and adaptations of native culture and loss of natural essence.

Socio-economic development

Rate of local units of high-growth companies, for the state as a whole.	**2,24**	%
Rate of local units of high-growth companies for agriculture, livestock, forestry, fishing and aquaculture.	**1,88**	%
Rate of local units of high-growth companies, for the Extractive Industries activity.	**3,4**	%
Survey of staff and their work		
Agriculture, livestock, forestry, fishing and aquaculture - Employed personnel.	24.566	Persons
Extractive industries - Persons employed	18.314	Persons
Manufacturing industries - Total occupied personnel	93.688	Persons
Electricity and gas - Employed persons	3.382	Persons
Water, sewerage, waste management and decontamination activities - Employed personnel.	4.265	Persons
Construction - Employed persons	89.253	Persons
Trade, repair of motor vehicles and motorbikes - Employed persons.	201.402	Persons
Transport, storage and mail - Employed persons.	42.317	Persons
Accommodation and food services - Employed persons	25.288	Persons

Information and communication - Employed persons	8.184	Person s

Figure 13 Table 7 - Socio-economic development - IBGE 2010

Socio-economic development data takes into account issues of local production, the number of workers in production and the number of existing companies. This information provides a relatively healthy outlook for the local economy, given that according to data from the national economy, there will be a significant slowdown in economic growth. This shows a decrease in the growth and expansion of activities, where the national GDP, which takes into account production factors, does not massively affect the region and local activities.

Financial, insurance and related services - Employed persons.	12.546	Person s
Real estate activities - Employed persons	3.692	People
Professional, scientific and technical activities - Employed persons.	15.541	People
Administrative activities and complementary services - Employed persons	61.847	People
Public administration, defence and social security - Employed persons.	289.000	People
Education - Employed persons	75.486	People
Human health and social services - Employed persons	44.610	People
Arts, culture, sport and recreation - Staff	4.214	People
employed.		
Other service activities - Employed persons	18.643	People

Figure 14 Table 8 - Employability in the state of Pará - IBGE 2010

The data we've analysed differs from the main human activities related to the whole of the state of Pará. Our biggest area of activity is administrative in general. It has been established that the state has a high concentration of public activities, such as town halls, defence and social security, which will have an impact on economic and social development rates, since many of these positions are qualitative leases compared to the others, providing their holders with a better quality of life.

Also according to the data, we have a lower rate of professionals in real estate and housing. This situation creates the prospect that there is not necessarily a great demand for housing in the state from these companies, probably due to the high cost of such housing or the current economic slowdown in the country.

EEZ

Ecological and Economic Zoning is an instrument of the National Environmental Policy, in accordance with item II of article 9º of Law No. 6.938/1981, and regulated by Federal Decree No. 4.297/2002, which aims to make socio-environmental development feasible in a way that favours and facilitates sustainable economic development. The state of Pará has 144 municipalities, 89 of which already have an EEZ.

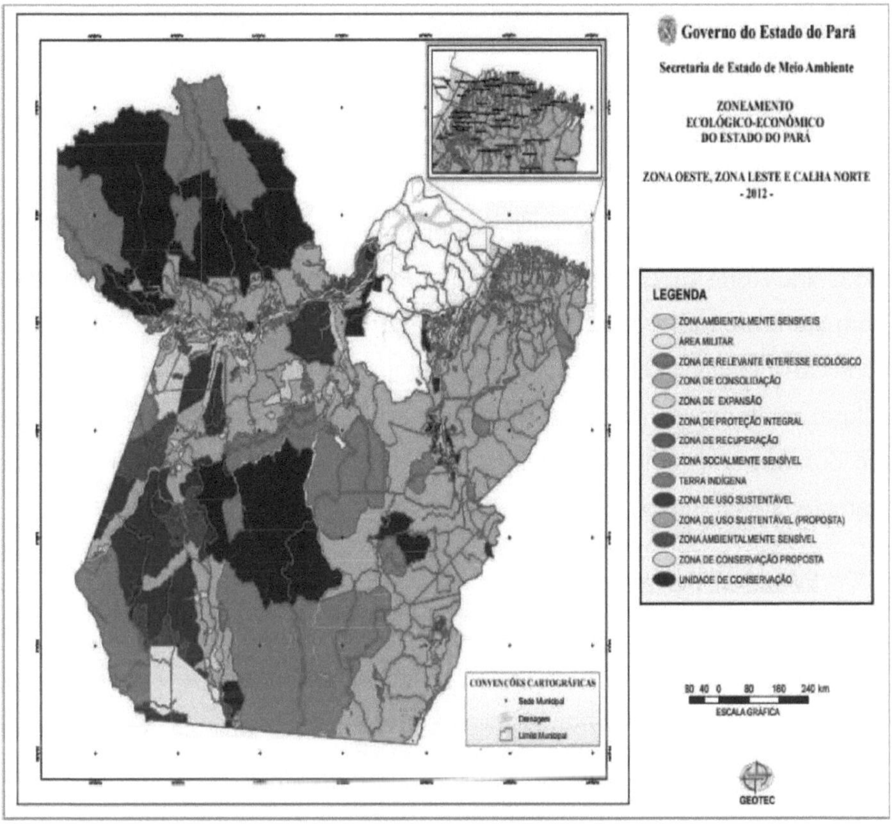

Figure 15 Map 5 - EEZ

Importance of Detailing

According to Sema, there are municipalities that need to be better studied due to their characteristics and particularities

and must be documented in order to become law. Because one region's proposal doesn't always fit the other. The information from the West is more focused on water resources, and the East is more based on livestock, agriculture and consolidated productive activities, and within this, there is a need to recover the 50 per cent of areas where there are no longer primary forests. Its recovery is of the utmost importance. This is how space is organised, according to the natural vocation and socio-economic condition of each area to be studied.[2]

East Zone and Calha Norte

With the approval by the Presidency of the Republic of the East Zone and Calha Norte EEZ, another stage of the EEZ in Pará has been completed. From there, we move on to more detailed municipal zoning. These will provide technical-scientific information on the biological, physical, socio-economic and cultural environment, so that each municipality can take hold of this knowledge and draw up the right public policies for mining, agriculture and fishing, for example, according to the type of soil.

West of Pará

The Western Pará EEZ project, which is part of the new development model for the Legal Amazon, represented by the Sustainable Amazon Plan (PAS), took just two years to complete. The proposal for Pará covers 19 municipalities under the area of influence of the BR-163 and BR-230 highways, a total of 334,450 square kilometres, which corresponds to 27% of Pará's territory. This area is 2.19 times larger than the state of Acre, 1.35 times larger than São Paulo and only 6% smaller than Germany. The EEZ presents the economic, environmental and social potential and limitations of the 19 municipalities in the region. As mentioned above, the state of Pará still has a small territorial area in terms of the EEZ.

---

2 http://www.semas.pa.gov.br/servicos/zee/importancia-do-detalhamento/ GENERAL INFORMATION ON ECOLOGICAL-ECONOMIC ZONING

Next EEZs Pará Coastal Zone

The EEZ for the Pará Coastal Zone, which involves 39 municipalities, is currently being drawn up by the state. Sector 1 of the Marajó Archipelago is made up of 12 municipalities (excluding the micro-region of Portei), made up 100 per cent of islands, including the micro-region of Furo de Breves and Arari. In the other EEZs, this archipelago is considered to be a single sustainable use zone. Now, on a smaller scale, each of the municipalities will be studied and divided according to their realities: forest areas, water resources. In the west of the archipelago there's Breves, which is one reality; Anajás is another reality, as it's more forested. In Afuá and Chaves, we can work with sea fishing and marine fauna. Sector 2 of the Coastal Zone, the Estuarine Zone, includes the Belém Metropolitan Region. Sector 3 is the north-east, the Salgado Paraense, where Salinas, Bragança, Maracanã, Marapanim, Viseu are located. The macro conservation units of the state of Pará were a direct indication of the Economic Ecological Macrozoning of the state of Pará, which became law in 2005. This macro-zoning used the work of the former IDESP, which carried out a general survey of the state of Pará to take to Rio92. At the time, the team surveyed the areas to be protected in the state in the form of Conservation Units for the ZEE, so the nature protection UCs are the result of the economic ecological macro-zoning of the Legal Amazon. From then on, work began on the detailed EEZ and on smaller units.

Calha Norte Conservation Units

The Calha Norte area covers 27 million hectares, 22 million hectares in Conservation Units alone, and Indigenous Lands (TI) account for more than 25 per cent of Pará.

Calha Norte - In addition to the tourist potential of the region, centred in the west of Pará and shaped by the waters of the Amazon River, another treasure that attracts the attention and concern of scholars is its timber potential.

Rural Environmental Registration

The Ministry of the Environment reports that the Rural Environmental Registry (CAR) is a fundamental tool for helping with the process of environmental regularisation of rural properties and possessions. It consists of collecting georeferenced information on the property, delimiting Permanent Protection Areas (APP), Legal Reserves (RL), remnants of native vegetation, consolidated rural areas, areas of social interest and public utility, with the aim of drawing up a digital map from which the values of the areas are calculated for environmental diagnosis, according to Semas there were around 133,710,000 registrations by 2014.710,000 registrations, taking into account the data collected by the IBG, and the rural population of 534,341,000 families in rural areas in 2010, suggest that this number has increased significantly, which is of great importance for traditional families in the region, as it is a way of guaranteeing the possibility of financing and contributing to the planning of public policies.

An important tool for assisting in the planning of rural property and the recovery of degraded areas, the CAR encourages the formation of ecological corridors and the conservation of other natural resources, helping to improve environmental quality, and is currently used by state and federal governments.

Green Municipalities

According to data obtained from SEMAS, there are 89 municipalities that have signed the Term of Commitment with the Federal Public Prosecutor's Office to become Green Municipalities. To this end, these municipalities use the ZEE to base themselves on the particularities of each region, and those that are in the macro-zone of consolidation of productive activities will have to prioritise work on the recovery of plant and aquatic ecosystems, according to the results and technical-scientific studies of these municipalities. The aim is to meet the needs of people in the state of Pará to improve their quality of life.

Survey of Forestry in the State of Pará - Major Productions

Vegetable Extraction and Forestry 2013 Quantities Produced		
Açaí quantity produced	111.073	Ton.
Cashew nuts	54	Ton.
Brazil nuts	9.023	Ton.

Figure 16 Table 9 - Forestry extraction in tonnes - IBGE 2013

Vegetable Extraction and Forestry 2013/ Production Value		
Acai	291.991	Thousand R$
Cashew nuts	154	Thousand R$
Brazil nuts	15.864	Thousand R$

Figure 17 Table 10- Extraction from Forestry in Reals - IBGE 2013

When analysing the extraction of permanent forestry, Açaí is the plant that predominates both in terms of production and value.

Miscellaneous Plant Extraction - Quantity Produced		
Jaborandi (leaf)	34	Ton.
Hevea (coagulated latex)	116	Ton.
Buriti (fibres)	257	Ton.
Miscellaneous plant extraction - Production value		
Jaborandi (leaf)	241	Thousand R$
Hevea (coagulated latex)	345	Thousand R$
Buriti (fibres)	326	Thousand R$

Figure 18 Table 11 - Extraction from Permanent Forestry - IBGE 2013

Given the data obtained, it was observed that the highest production is not always the one that generates the greatest economic value, as is the case when comparing Buriti production, which is the highest of the three crops compared, but is not the one that obtains the greatest economic value.

Survey of permanent and temporary agriculture in the state of Pará

The information in the tables below shows that beans and manioc are the main production items found in the state of Pará. The figures differ on the production capacity of manioc and the space in ha used for production. used in production. Aspects involving the monetary return on each item produced are also taken into account, and again the return on cassava outweighs the other crops.

From the tables below, it can be seen that the main issue of high cassava production is proportionally related to the amount of land available to produce this food, and the market value drives supply and demand both locally and for exports.

PLANTED/TEMPORARY AREA		
Cassava	302.300	ha
Pineapple	10.777	ha
Peanuts	112	ha
Beans	43.941	ha
Sweet potatoes	30	ha

Figure 19 Table 12 - Temporary Agriculture hectares - IBGE 2013

PRODUCTION OF TEMPORARY CROPS		
Sweet potatoes	150	Tonnes
Beans	30.737	Tonnes
Peanuts	147	Tonnes
Cassava	147	Tonnes

Figure 20 Table 13 - Temporary Agriculture in tonnes - IBGE 2013

PRODUCTION/TEMPORARY VALUES		
Sweet potatoes	155	A thousand reals
Beans	65.493	A thousand reals
Peanuts	232	A thousand reals
Cassava	2.208.029	A thousand reals
Pineapple	263.636	A thousand reals

Figure 21 Table 14- Temporary production values in Reais - IBGE 2013

In this comparison of temporary crops, cassava predominates in all three respects: greater area, greater production and greater market value. But if you analyse the other crop options, you can see that pineapple is a promising crop, as it requires a smaller area and acquires considerable value.

Table - produced using IBGE/2013 data

PLANTED/PERMANENT AREA		
Cocoa	97.176	ha.

Banana	43.510	ha.
Dendê	54.475	ha.
Pepper	13.948	ha.

Figure 22 Table 15 - Permanent Agriculture in hectares - IBGE 2013

PERMANENT CROP PRODUCTION		
Cocoa	79.727	ha.
Banana	585.943	ha.
Dendê	1.040.5 38	ha.
Black pepper	30.885	ha.

Figure 23 Table 16- Permanent Agriculture in tonnes - IBGE 2013

PRODUCTION VALUES / PERMANENT		
Cocoa	336.84 8	Thousa nd R$
Banana	411.92 6	Thousa nd R$
Dendê	263.28 5	Thousa nd R$
Black pepper	365.33 4	Thousa nd R$

Figure 24 Table 17 - Permanent Agriculture in Reais - IBGE 2013

In this comparison of permanent crops, there is a discrepancy in the three aspects mentioned. The smallest area, the least productive and the most valuable is chilli pepper, which is the predominant crop in the state of Pará, second only to bananas in value. However, other crops are also economically important.

Economic Investment

Loans made by farmers in 2012 throughout the state, according to the MDA source in the Rural Credit Yearbook, of which Agricultural 1,766 - Contracts and Livestock 319,972,670 - Contracts.

Crops that act as natural pesticides

Andiroba - Andiroba *(Carapa guianensis) is* a medicinal plant also known as andirova, angirova, carapa, andiroba-branca, carapinha, camaçari, castanha-mineira, among other popular names. It belongs to the Meliaceae family. Indigenous Amazonian tribes use the bark and leaves of the andiroba tree as a tea to relieve fevers and intestinal worms, as well as applying the tea topically to ulcers, insect bites, skin and scalp parasites and other skin problems. To aid digestion, the bark is soaked in water for a day and a cup is taken before meals. In Brazil, the fruit's oils are used internally to treat coughs and applied topically to remove ticks from animals.

Neem - NEEM is a tree native to India, the Burna region and the arid areas of the Indian subcontinent and south-west Asia.

In these regions, NEEM is used as a medicinal plant of great importance due to its positive effects on the health of crops, animals and man himself, as it has been cultivated for many decades in the arid zones of Asia - in India, approximately 18 million NEEM trees grow there today. The tree has been taken to other hot regions of the planet, such as West Africa, as it adapts well to humid tropical zones with an altitude of 800 metres, as well as providing shade.

NEEM arrived in Brazil in 1993, on the initiative of the Brazilian Research Company - EMBRAPA, with a view to viability due to its rapid growth and exceptional resistance to drought, which makes it favourable to Brazil's tropical climate.

In North America and Europe, NEEM doesn't develop because of the cold, which is also the case in Brazil, in the southern states.

Crotalaria - This is a legume that grows very little, reaching a maximum of 60 cm to 1 metre in height, and does not grow as tall as the others, with many species reaching up to 4 metres in height. Many of the common crotalarias only flower after 6 months or 1 year. Crotalaria *breviflora*, on the other hand, flowers straight away. It can even be planted on housing estates. It can be grown as mulch between the rows of perennial crops, especially to combat the "Dengue Mosquito" - Aedes aegypti. If well cared for, it flowers in up to 90 days, which is 4 to 5 times faster than the other 550 species of crotalarias in the world, which generally take 6 to 1 year to flower, this species of *breviflora* is not as tall as the other existing species. Due to its relatively small size (dwarf), its non-climbing habit and the fact that it is a host species for nematodes, it flowers quickly, making it one of the best "green manures" to be used between the rows of coffee plantations or orchards.

Chilli - Chilli is a heat-demanding plant, sensitive to low temperatures and intolerant of frost, so it should preferably be grown in the high temperature months, a condition that favours germination, development and fruiting, thus obtaining a product of high commercial value at a lower production cost.

For pepper plants, the ideal average monthly temperatures are between 21°C and 30°C, with the ideal average minimum temperature being 18°C and maximum temperatures around 35°C. Temperatures above 35°C are detrimental to fruit formation. Germination is favoured by suitable soil temperatures between 25°C and 30°C, with 30°C being the temperature at which the shortest interval of days between sowing and germination occurs, and soil temperatures of 10°C or less inhibit germination. For seedlings, the best growth is achieved at temperatures between 26°C and 30°C, with 27°C considered the ideal temperature to favour plant development.

Low temperatures make production unfeasible, causing flowers and fruit to fall, as well as negatively influencing the pungency and colour of the fruit, leading to a reduction in commercial value, especially if the product is destined for industrialisation. These facts are easily verified when comparing plantings under plastic protection and in the open field. In addition to these effects, low temperatures can also cause mature leaves to become stunted, young parts to wilt and slow growth. There is no information on the effects of photoperiodism or thermoperiodism on pepper plants.

CHAPTER 3

SURVEY ON OIL PALM CULTIVATION

The product - Palm oil

The oil palm, whose scientific name is Elaeais *guineensis Jaqquim*, is of African origin. The oil extracted from this palm is palm oil. Today it is widely used in Northeastern cuisine, but its current use has gradually expanded into other sectors such as cosmetics and biodiesel. Studies show that its consumption has existed for over 5,000 years and its use in Brazil began with the arrival of the slaves (Souza 2000). There has been a tendency for oil palm cultivation to expand in the north of Brazil and in the north-east, but the north is the region where most of the food is produced.

Among oilseeds, oil palm has the highest productivity, reaching up to 8 tonnes of oil/ha/year, surpassing soya, castor beans and maize (Teixeira, EMBRAPA).[3]

Oil palm cultivation has expanded in this region due to the climatic conditions and the quality of the soil, which favours the quality of the end product and its by-products, such as oils, biodiesel and other items. The most important climatic factors for growing oil palm are rain, hours of sunshine and high temperatures (Souza 2000). Although it is grown in different types of soil, variations in physical and chemical properties cause significant differences in production. The most important parameters are an effective depth of more than 90 cm, a loamy or clayey texture, a strong or moderate structure, moderate permeability, flat or gently undulating terrain, not stony, without iron, aluminium or manganese concretions and without a dense, consistent layer (Souza 2000).

---

[3] *http://www.redebiodiesel.com.br/projeto/dende - Technological development of oil palm for the production of biofuels -Paulo César Teixeira - Embrapa Amazônia Ocidental*

Rainfall is another important factor in the oil palm production cycle, as rainfall of over 100 mm can ensure attractive production and fulfil the soil's needs for this type of plant. The incidence of sunlight for this species also influences its growth and productivity, as it must get at least 5 hours a day to obtain the nutrients it needs to survive. Temperature determines growth and production, and it has been observed that the highest yields are obtained in regions with small temperature variations and where the annual average is between 25 and 27 degrees centigrade with no occurrence of minimum temperatures below 17°C for prolonged periods (Souza 2000).

There are three varieties of oil palm: Dura, Psífera and Tenera. The Dura has a husk more than 2 mm thick and fibres in the pulp; this variety is used as a female plant in the production of commercial hybrids. The Psífera has fruit without a husk separating the pulp from the kernel. It is used as a pollen supplier in the production of commercial hybrids. Tenera has a husk thickness of less than 2 mm and a fibrous ring around it. It is obtained by crossing the Dura and Psifera varieties and is recommended for commercial planting (Souza 2000).

Oil palm cultivation requires a large amount of fertiliser to obtain high yields. The palm requires around 192.5 kg of nitrogen, 26 kg of phosphorus, 251 kg of potassium, 61.31 kg of magnesium and 99.3 kg of calcium per hectare per year to grow and produce 25 tonnes of bunches per hectare per year. However, in order to use it in this way, the soil must be analysed and corrected accordingly (Souza 2000).

Pests - one of the main pests of the oil palm of economic importance is Rhynchophorus palmarum, whose larvae feed on the tissues of the stipe, making galleries that can cause internal rot; when it reaches the meristem it causes the death of the plant. Rhynchophorus palmarum is the main vector of the nematode that causes red ring disease. Other pests occur sporadically, such as the root borer caused by Saglassa valida and defoliator caterpillars such as Sibine fusca and Brassolis sophoraea. Fusariosis, caused by the fungus Fusarium oxysporun, marchitez, caused by the protozoan Phytomonas sp and fatal yellowing whose causal agent is unknown (Souza 2000). [4]

Production

Harvesting begins in the third year after planting, with a yield of between 6 and 8 tonnes of bunches per hectare, reaching its maximum yield in the eighth year, when it can reach 25 tonnes of bunches per hectare, a yield that remains at this level until the 17th year, declining slightly until the end of its productive life, which is around 25 years. Two types of oil can be extracted from oil palm fruits: pulp oil, known in Brazil as palm oil, and palm kernel oil. The oil yield is 22% of the weight of the bunches for pulp oil and 2% for palm kernel oil (Souza 2000).

[4] http://www.ceplac.gov.br/radar/dende.htm - Executive Committee of the Ploughing Plan

CHAPTER 4

BIODIESEL

Biodiesel in the world

The world currently produces much less biodiesel than ethanol, but unlike alcohol, whose production is concentrated in Brazil and the USA, biodiesel production is spread across several countries around the world, with their own legislation and commercial strategy defined in each one. This proliferation across the continents makes it easier for the product to enter the list of world commodities, favouring the development of its market on a global scale.

With the rapid evolution of its biodiesel programme, Brazil is well positioned on the world stage. The country made it compulsory by law to use 2 per cent of the product in diesel in 2008, and has always raised the minimum percentages earlier than imagined, reaching the position of fourth largest producer in the world by the end of the first year of compulsory production. But the country is still behind three economic powerhouses - Germany, the US and France - which started developing biofuel earlier.

The economic bloc with the largest industrial and technological park for this green fuel is precisely the European Union, home to some of the largest biodiesel programmes in the world. There, the addition of 5.75 per cent biodiesel will be compulsory from next year. That alone means a demand for more than 50 million litres of biodiesel a day. In addition to the mandatory use (which some countries have raised to even higher percentages), there is also an aggressive fiscal policy, with tax incentives and heavy taxation on oil derivatives.

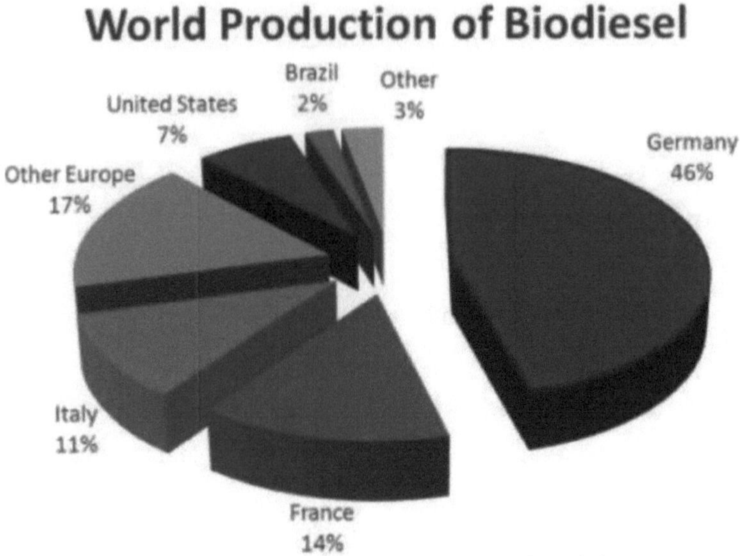

Figure 25 Source: Based on data from the European Biodiesel Commission.

Germany stands out the most. The world's largest producer of biodiesel, with 3.1 billion litres in 2008, the country is also a major consumer. With the mandatory 5% addition already in force, Germans consume 8.8 million litres of biodiesel a day. To meet this demand, they use canola or rapeseed, planted on a large scale to nitrogenise the soil for agriculture. Once harvested, the material becomes bran for animal feed and its oil is used to produce vegetable diesel[5] .

CO2 emissions: when biodiesel is burnt, the CO2 released into the atmosphere will be recycled by the growth of the plants that will later be used to produce more biodiesel. A study sponsored by the US Department of Energy and the US Department of Agriculture (USDOE/USDA) concluded that biodiesel reduces carbon dioxide emissions by 78 per cent when compared to petrodiesel.

---

[5] http://www.biodieselbr.com/biodiesel/mundo/biodiesel-no-mundo.htm

Carbon monoxide: carbon monoxide emissions are 48 per cent lower than diesel. Biodiesel has a higher oxygen content, which helps to achieve a higher combustion rate.

SOx emissions: biodiesel is basically sulphur-free. Therefore, sulphur oxides and sulphates are essentially eliminated compared to diesel.

Particulate matter: particulate matter emissions are approximately 47 per cent lower than total diesel emissions

Hydrocarbons: total hydrocarbon emissions are on average 67 per cent lower than in diesel.

Nitrogen oxides: NOx emissions increase when using pure biodiesel by around 10 per cent. However, some technologies and additives are being researched in order to reduce this problem[6] .

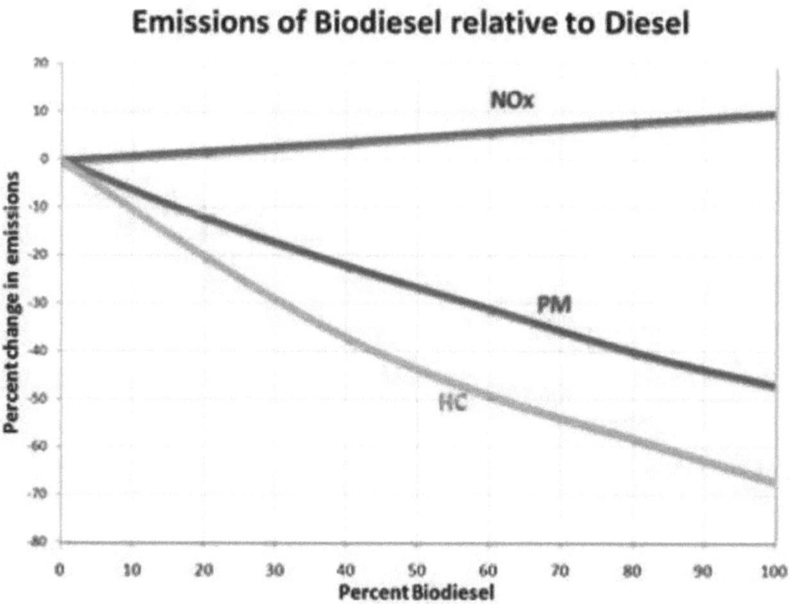

Figure 26 Emissions: Average emission impacts for heavy-duty engines (Data covers biodiesel from different crops).

Biodiesel in Brazil

In 2003, the first concrete studies began into the creation of a biodiesel

[6] http://www.esru.strath.ac.uk/EandE/Web_sites/06-07/Biodiesel/biodiesel2p.htm

policy in Brazil and, in December 2004, the government launched the National Programme for the Production and Use of Biodiesel (PNPB). The initial aim was to introduce biodiesel into the Brazilian energy matrix, with a focus on social inclusion and regional development. The main result of this
The first phase was the definition of a legal and regulatory framework, with the publication of two laws and various infra-legal normative acts.

In this way, the PNPB institutionalised the normative basis for the production and commercialisation of biodiesel in the country, involving the definition of the tax model for this new fuel and the development of mechanisms for the inclusion of family farming, embodied in the Social Fuel Seal. This work was guided by certain very clear guidelines: a policy of social inclusion; the utilisation of oilseeds in accordance with regional diversity; security of supply for the new fuel; a guarantee of quality for the consumer; and the search for competitiveness in relation to petroleum diesel.

Management of the PNPB is the responsibility of the Interministerial Biodiesel Executive Commission (CEIB), coordinated by the Civil House of the Presidency of the Republic, and also by the Management Group, coordinated by the Ministry of Mines and Energy (MME). The Interministerial Executive Committee (CEIB) is responsible for implementing and monitoring the programme, proposing normative acts, as well as analysing, evaluating and proposing other recommendations and actions, guidelines and public policies. The Management Group is responsible for carrying out operational and administrative management actions aimed at fulfilling the strategies and guidelines established by the CEIB.

Since the launch of the PNPB, the private sector has been contributing resources, investing in the distribution of the fuel, in laboratories, in research, in the production of raw materials, all thanks to the security of the regulatory environment provided by the definition of targets and the creation of a legal framework for biodiesel.

The blending of biodiesel into fossil diesel began in December 2004 on an experimental basis. In January 2008, the legally mandatory blend of 2% (B2) came into force throughout the country. As the Brazilian market matured, this percentage was successively increased by the National Energy Policy Council (CNPE) until it reached 5% (B5) in January 2010, three years ahead of the target set by Law No. 11,097 of 13 January 2005.

Biodiesel is regularly sold mixed with petroleum diesel at more than 30,000 petrol stations across the country. Several indicators confirm the success of the PNPB. Production of this type of biofuel jumped from 69 million litres in 2006 to 2.7 billion litres in 2011. In 2012 it should reach 2.8 billion litres. This makes Brazil one of the world's biggest biodiesel markets, along with Germany and the United States, which have been producing and consuming this renewable fuel for much longer. Other important markets are France, Spain, Italy and Argentina.

Also noteworthy is the rapid evolution of industrial biodiesel production capacity. At the end of 2011, 56 units were authorised to produce and market the biofuel, with a total nominal capacity of 6 billion litres/year. The participation of small farmers is also significant. Of this industrial capacity, around 78% (4.7 billion litres/year) comes from plants that hold the Social Fuel Seal, a certificate provided by the government to production units that meet the requirements for including family farming in the biodiesel production chain.

Since the launch of the PNPB until the end of 2011, Brazil has produced 8.3 billion litres of biodiesel, reduced diesel imports by 5.3 billion dollars and made a positive contribution to the Brazilian trade balance.

The PNPB's basic guiding principles of promoting social inclusion and regional development have been pursued continuously, with important results achieved and lessons learnt. The Programme is supported by resources for research, development and innovation (RD&I) throughout the production chain, from the agricultural phase to industrial production processes, including by-products and storage. The current tax model gives Brazilian biodiesel the unique characteristic in the world of a biofuel supported by socially orientated public policies.

Table 27: Bitidiesel Chain Tax Regime

Federal taxes	Siodrosol				Petrol Diesel
	Family farming in the north, north-east and semi-arid regions with castor bean or oil palm	Family Farming	North, NE and If ml 'ăr Ido with castor bean or palm	General rule	
IPI	Al [zero quota	Allquota zero	Allquota zero	Allquota zero	Allquola zero
CIDE	Nonexistent	Nonexistent	non-existent	Nonexistent	RS 0.07
PI&COFINS	100% reduction	68% reduction	31 per cent reduction	RS 0.22	RS 0.15
Total Federal Taxes	RSrtitno R$-	RS/litre R$ 0.07	R$/IHro RS 0.15	RS/lil RS 0.22	RS/litre RS 0.22

Source: COSTA *ei al.*, 2008

Figure 27 Biodiesel chain tax regime

Also noteworthy is the inclusion of family farming in the biodiesel production chain through the Social Fuel Seal, granted by the Ministry of Agrarian Development (MDA). The Social Fuel Seal is a certificate awarded to biodiesel producers who purchase a minimum percentage of raw materials from family farmers; sign contracts with family farmers, establishing deadlines and conditions for delivery of the raw materials; and provide technical assistance to farmers[7] .

Cartogram of nominal capacity and biodiesel production (B100). By region thousand^3 /year - 2013 in Brazil

[7] http://www.mme.gov.br/programas/biodiesel/menu/biodiesel/pnpb.html

Figuro 28 Souroo: ANP

Composition and use

Palm oil contains equal proportions of saturated fatty acids (palmitic 44% and stearic 5%) and unsaturated fatty acids (oleic 40% and linoleic 10%). It is a natural powerhouse of vitamin E, tocopherols and tecotrienois, which act as antioxidants. It is also rich in beta-carotene, an important source of vitamin A.

It is the most suitable oil for making margarine because of its consistency and because it does not branch out. It is excellent as a cooking and frying oil, and is also used in the production of vegetable butter (shortening), which is suitable for making bread, cakes, pies, fine biscuits, creams, etc. The main non-edible use of palm oil is as a raw material in the manufacture of soaps, washing powders, detergents and biodegradable fabric softeners, and it can also be used as fuel in diesel engines (Souza 2000).

Product production stages

Immediately after harvesting, the following steps must be followed:

Sterilisation - the purpose of this is to inactivate enzymes that cause acidity, make it easier to detach the fruit from the bunches and cause the oil-containing cells to rupture.

Threshing - the purpose of which is to separate the fruit from the bunch

Digestion - breaks down the structure of the pulp cells, facilitating the release of oil and pressing.

Pressing - the mass coming out of the digester is pressed, separating the oil and a mixture of fibres and seeds, which then passes through the shredder, which separates the fibres from the seeds by means of ventilation.

Utilisation of by-products

The fibres are used as fuel in boilers;

The seeds are transported to the dryers and after drying they are sent to the breakers and then the shells and kernels are separated which, after crushing, still extract the palm kernel oil;

The remaining residue is the cake, which contains 14 to 18 per cent protein and can be used as a component of animal feed (Souza 2000).

Research and innovation

The Embrapa Agro Energia team dedicated to researching processes for obtaining agro-energy products is studying the production of biodiesel from palm oil. According to researcher Itânia Soares, this oil is usually more acidic than soya oil. Because of this, the conditions for the chemical reaction that produces biodiesel (transesterification) are different, which implies the need for adjustments. "Our studies are aimed at reducing costs, stages and the generation of effluents in the production of palm biodiesel," he specifies. The development of a quality control methodology for the diesel/palm biodiesel blend is another research objective (Chies 2012).

With a view to making the most of the biomass from the oil palm, uses are also being studied for the empty bunches, known as stalks. For every tonne of palm oil, around 1.1 tonnes of this waste is generated. Utilising it would add value to the production chain and prevent it from becoming an environmental liability. In the opinion of researcher Leonardo Valadares, "the full utilisation of oil palm is essential for the sustainability of the business". (Chies 2102).

The stalks are rich in cellulose. Embrapa's work consists precisely of extracting cellulose nano-fibres from them, which can be used for various purposes, including paper production and reinforcing natural rubber. "Cellulose is a biodegradable material with properties that make it useful in various applications," emphasises Valadares (Chies 2012).

Embrapa Agro Energia is also studying POME, an effluent generated in the oil extraction process. According to some research, it is rich in nutrients and is currently used in the fertigation of oil palm plantations (Chies 2012).

Palm-based cellulose

Research led by Embrapa Agro Energia (Brasília/DF) has succeeded in purifying cellulose from empty bunches of oil palm, popularly known as dendê. In addition, the scientists extracted cellulose nanofibres from this material. The results obtained are expected to contribute to the development of applications with greater added value for palm oil production waste.

Currently, industries obtain pulp mainly from planted forests and use this pulp to produce paper. The material obtained in Embrapa Agro Energia's laboratories, however, could have other applications. The nanofibres are being characterised using various instrumental techniques and, after this stage, they will be tested as a reinforcement for natural rubber. The idea is to obtain a material that can be incorporated into rubber without compromising its properties.

Embrapa Agroenergia researcher Leonardo Valadares, who is coordinating the research, explains that nanoparticles have different properties that allow them to generate unique materials. "New technologies can be generated to replace petrol derivatives, for example," he says. The research project using empty oil palm bunches is part of the AgroNano Network, a nanotechnology programme coordinated by Embrapa Instrumentação (São Carlos/SP). The network has already studied the preparation of nanocomposites made of natural rubber and cotton fibre.

Oil palm is one of the most important sources of vegetable oil in the world. There are great expectations for an increase in oil palm production in Brazil, mainly to supply raw materials to biodiesel plants, in the context of the National Programme for the Production and Use of Biodiesel. The agro-ecological zoning of the crop carried out by Embrapa and the Ministry of Agriculture, Livestock and Supply shows that there are around 30 million hectares of land in the country suitable for oil palm plantations, without compromising environmental protection areas.

Expanding the planted area, however, would lead to an increase in the volume of waste generated. For every kilo of palm oil obtained, around 1.1 kilos of empty bunches are left over. These are currently returned to the fields to fertilise the plantation. However, utilising the cellulose contained in this material could generate products with high added value.

Cellulose nanofibres from oil palm bunches were obtained on a bench scale. One of the research challenges now is to increase the scale of production. In addition to Embrapa Agroenergia and Embrapa Instrumentação, Embrapa Cerrados and Embrapa Recursos Genéticos e Biotecnologia, both in Brasilia/DF, are taking part in the project[8] .

Brazil's palm oil production region

IBGE and CONAB surveys show that oil palm is grown in the North,

---

[8] Source: https://www.embrapa.br/busca-de-noticias/-/noticia/1497582/nanofibras-de- cellulose-can-be-extracted-from-dende-residues

Northeast and Centre-West regions, but the predominant region is the North. The states of Pará, Amapá and Bahia stand out for their production.

Palm oil in the North

Abundant in palm species and suitable for growing the exotic species Elaeis *guineenses* (oil palm), the northern region is characterised by its territorial extension, vast existing forest cover, humid and equatorial climate (in a large part of its area) and high rainfall rates. In the region, the state of Pará stands out as the largest producer of palm oil, producing 100,000 tonnes in an area of 50,000 hectares. According to Campos (2003), this state has 2 million hectares of unproductive land with a climate and soil suitable for oil palm, with all the infrastructure, including a port and electricity, at a maximum distance of 200 km from Belém.

The productivity of palm/oil palm in the Amazon Region is over 5.0 tonnes of oil per hectare/year and it is also possible to use other native oleaginous species to supply small industrial units, which would contribute to solving the energy deficit in the North Region, especially in communities that are practically isolated and difficult to access (2006 STCP Engenharia de Projetos Ltda.).

Oil Palm Zoning in Brazil

Its aim is to regulate the expansion of palm oil production in Brazil and provide instruments to guarantee production on an environmentally and socially sustainable basis, the Agroecological Zoning of Oil Palm Decree ZAE delimited only suitable areas (soil and climate) in anthropised regions without environmental restrictions Bill Prohibiting the suppression of native vegetation throughout the national territory as stated on the MMA website.

Pará is the state's main producing region, with 37 municipalities suitable for the activity, according to the Agroecological Zoning of Oil Palm. Currently, according to data from the Pará State Department of Agriculture (Sagri), adding the areas owned by the business sector to those occupied by oil palm in family farming, the region accounts for around 166,000 hectares occupied by the crop.[9]

The areas considered in the agro-ecological zoning of oil palm are those in the national territory with low climatic risk and are divided into two large groups:

1. The suitable anthropised areas in the states of Acre, Amazonas, Amapá, Maranhão, Mato Grosso, Pará, Rondônia and Roraima;
2. The suitable anthropised areas in the states of Alagoas, Bahia, Espírito Santo, Pernambuco, Rio de Janeiro and Sergipe. From this total, the areas with native vegetation, protected areas (protected areas) and the areas with native vegetation (protected areas) were subtracted from the zoning.

   conservation and indigenous lands), urban areas and public easements.

This law is important for limiting the use of oil palm cultivation, without deforestation and invasion of indigenous lands and preservation areas.

Main problems

The positive and negative aspects that are intensely linked to the proposed theme, the expansion of oil palm, have had a lot of repercussions and are currently one of the biggest problems for small farmers in the state of Pará.

The study of oil palm took on greater proportions in the mid-2000s when the possibilities of using oil palm for biodiesel production were discovered. According to climatological requirements, the north and northeast are the regions that predominate in this type of crop, with greater production in the north of Brazil.

As this crop is guaranteed to make a profit, the big industries are investing more and more in this production, so they are leasing land in Pará for this cultivation. As oil palm takes around 4 to 5 years to produce its first crop,

[9] http://reporterbrasil.org.br/documentos/Dende2013.pdf - 3 Created in 2007, Pronaf Eco, designed to stimulate the production of agro-energy crops by family farmers as part of the National Programme for the Production and Use of Biodiesel (PNPB), has been an agricultural public policy tool in the north-east of Pará since 2010. It currently provides credit of up to R$80,000 to family farmers integrated with large companies.

small farmers are at the mercy of this crop, becoming employees of the big industries to look after the product. The salary is not enough to support a family with all the basic necessities. The areas that were once used for subsistence farming are now mostly used for oil palm cultivation.

Many families leave their land in search of better work opportunities, which leads to a rural exodus that makes room for monoculture and the monopoly of large agro-industries, much like what happened with soya.

With monoculture, the negative impacts are greater, as environmental degradation extends to the soil, water, biodiversity, fauna and flora with the use of more chemical fertilisers and pesticides, as monoculture limits natural predators.

There are many government plans to support small-scale farmers, but little of them are utilised, so they end up being disadvantaged by outdated legislation. For consistent changes in the current scenario and demands in agriculture, processes would need to be reviewed, as would legislation on mass exploitation and monocultures. Many actions can be taken to change this scenario in order to meet the sphere of needs by balancing the various sectors involved. Relevant legislation to be used in assigning proposals.

Land Statute	Guidance on leasing land	LAW No. 4.504, of 30 November 1964.
Pronaf National Family Farming	There are several types of credit line and they all cover the suggestions mentioned.	Law No. 4.829, of 5 November 1965 - Establishes the National Rural Credit System (SCNR); CMN (legislates) and BACEN (publishes); - Instituted the Rural Credit Manual - MCR (has the force of law);

		Decree No. 1.946, of 28 June 1996, creates the Nation al Programme of Streng thening Family Farming -.
Embrapa Brazilian livestock and agricultural research company	Research and training support	Authority subsidises Guid elines cultivation techniques
Conab National Supply	Buy commercialisation of crops	Law No. 8.029, of 12 April 1990, having started its activities on 1° January 1991.
PAA (Food Acquisition Programme)	Family Farming Plan Responsible for implementation of programme, whose guidelines established published in Resolutions.	0 PAA is an instrument of public policy instituted by Article 19 of Law No. 10.696, of 2 July 2003, and regulated by Decree No. 4.772, of 2 July 2003, which was amended by Decree No. 5.873, of 15 August 2006.
EEZ Ecological Zoning Economic	Sustainable use of natural resources and balance existing ecosystems, making sustainable	Law No. 5.887, of 9 May 1995 /Decree No. 1.148, of 17 July 2008 Normative Instruction No. 37/2010, of 2nd February

	development based on making development compatible	2010/construction
	socio-economic and environmental conservation, balancing the ecosystems. An indispensable tool for the development of oil palm.	Normative No. 39/2010, of 04 February 2010.
ZAE - Dendê 0zoning is a fundamental instrument of technical orientation scientific sustainable implementation of the palm oil production chain.	Approves the agro-ecological zoning of oil palm cultivation and provides for the establishment by the National Monetary Council of rules relating to financing operations for the oil palm segment, under the terms of zoning.	Decree No. 7.172 of May 2010
	oil palm< zoning.	), in accordance with
Forest Code Brazilian	It is the law that establishes the general rules on where and how the native	(Federal Law No. 12.651/2012) establishes a five-year deadline for all states to draw up and

	vegetation of Brazilian territory can be exploited , determining which areas must be preserved and which regions are authorised to be used for various types of rural production.	approve their EEZs, according to a unified methodology established in federal standard.
CAR Rural Environmental Registry Integrated tool of the Brazilian Forest Code	An important tool to help plan rural properties and recover degraded areas, the CAR promotes the Formation of ecological corridors and the conservation of Other resources contributing to the improvement of	> Law No. 5.887, of 9 May 1995. Decree No. 1.148, of 17 May 1995
	environmental quality, being currently used by state and federal governments.	July 2008. > Normative Instruction No. 37/201 0, of 2 February 2010. > Normative Instruction No. 39/201 0, of 4 February 2010.
Co-operatives	A co-operative is a civil society made up of at	Law 5.764 of 1971

	least 20 people,	
	managed in a democratic and participatory way, with common economic and social objectives.	

Figure 29 Table 18- Legislation applied

Dendê's main by-products and their uses

Dendê Products	By-products	Benefits
Oil palm cultivation	Carbon fixation in the soil	Richer soil
Palm oil (pulp)	2 types of oil Stearin Olein Food use, cosmetic use, biodiesel	Guaranteed market is the most used oil in the world
Palm kernel oil	Food use, cosmetic use, biodiesel.	Agri-energy safety and benefits
Pulp fibre	Feed, fertiliser	Organic fertiliser market
Bucha (Rest of the bunch)	Fertiliser, biomass	Organic fertilisation, biomass.
Liquid Smoke	Smoked	smoked food products use this by-product
Straw	Handicrafts	Household utilities
Bark	Charcoal	Market
Waste	2nd generation ethanol o	Automotive energy

Figure 30 Table 19- Dendê utilisation (Caires, Santos 2015).

Representative diagram of the Biodiesel extraction processes

The biodiesel production process, starting from any greasy raw material, involves the operational steps shown in the figure.

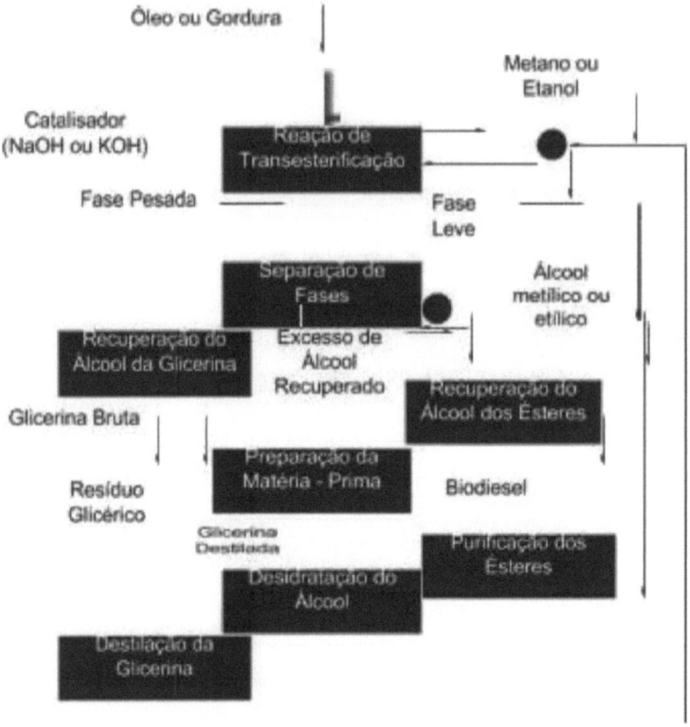

Figure 31 Table 20 - Representative diagram of palm oil extraction

The process of producing biodiesel from oil palm after harvesting must be quick, no more than 24 hours after harvesting, thus avoiding an increase in acidity. Therefore, as presented by Embrapa researchers, the acidity of oil palm must be controlled so as not to jeopardise the process of obtaining the by-product[10] .

Evaluation of the oil palm life cycle

Life Cycle Assessment (LCA) is a study that measures the environmental impacts associated with a product (or a production system), comprising the

[10]h ttps://www.youtube.com/watch?v=SOCY6KAtAMY
FAPEAM: Research into palm oil/25 **January** 20110 Dendê oil, a typical product of Brazilian cuisine, is now one of the most commercialised oils in the world, to be used in a variety of applications, from cooking to biofuels.
In this edition of "Amazonas Does Science", Embrapa Amazônia Ocidental's work with Dendê oil is described by researcher Vanderlei Alves de Lima.

stages from the removal from nature of the raw materials used in production to the disposal of the product in landfills, its reuse or recycling. In the case of palm biodiesel, it is from cultivation to the use of the product to produce fuel, including all the intermediate stages (harvesting, transporting the seeds, extracting and processing the oil, distributing the oil for consumption, disposing of waste, etc.). LCA generates a wide range of results, which generate recommendations that can contribute to more sustainable production.

In the LCA of palm oil, the focus was on the Global Warming Potential (GWP) and the main objective was to verify whether there was a balance between the emission of greenhouse gases during the production of palm oil in the Amazon region and the absorption of these gases by oil palm trees, which promotes plant growth. For this study, the ideal conditions for oil palm production were considered. The oil palm farm studied is located on abandoned land in the state of Pará, previously occupied by agricultural crops and some pastures. The characteristics of the region's soil and climate provide the highest productivity of this fruit in Brazil. The life cycle of the oil palm goes from the production of seedlings in nurseries to the cutting down of the adult palm trees, at around 30 years of age.[11]

[11] **http://www.canalciencia.ibict.br/pesquisa/0273-Avaliacao-do-ciclo-de-vida-da- biodiesel-production-from-dende.html**

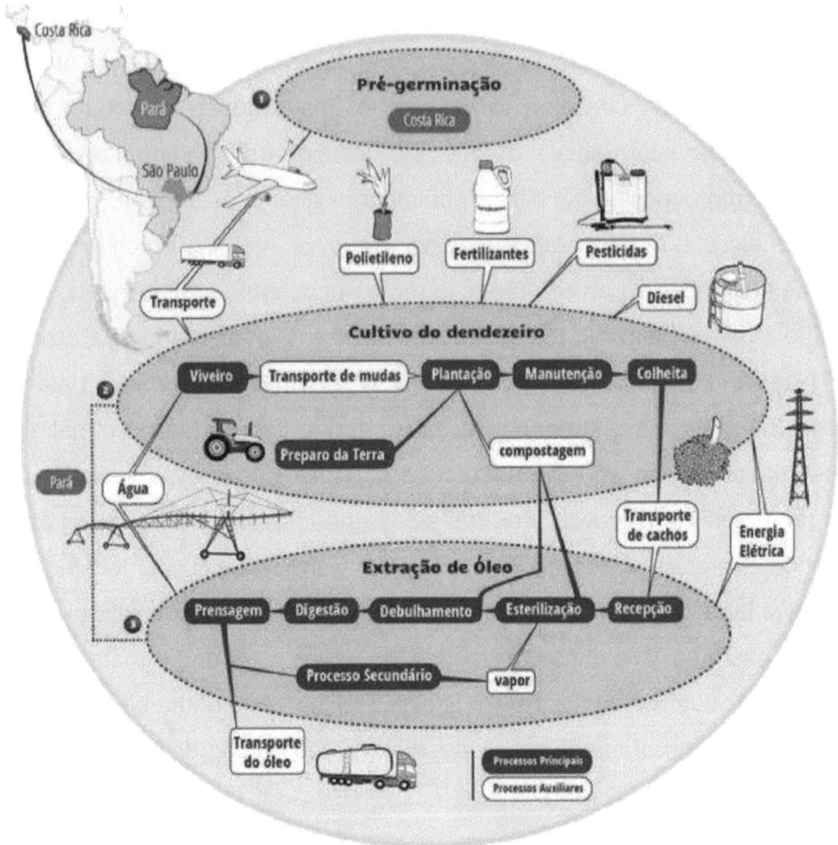

Figure 32 Figure 1. Life cycle of palm oil biodiesel production. 1. After pre-germination in Costa Rica, the seedlings are transported by plane to São Paulo and from São Paulo to Belém. They are then transported by lorry from Belém to the planting region.

The palm oil life cycle inventory accounts for all the raw materials, products, by-products and waste generated and consumed from growing the seedlings to extracting the oil. The table below shows the quantities of raw materials needed, the products and by-products obtained during the cultivation phase (table 1) and production (table 2) of 1 tonne of palm oil[12] .

Processes	Entry	Quantity	Output	Quantity
Reception	Bunch of fresh fruit	5000 kg	Palm oil	1000kg
Sterilisation	Diesel	2.96kg	Pairnlste oil	75kg

[12] **http://www.canalciencia.ibict.br/pesquisa/0273-Avaliacao-do-ciclo-de-vida-da- biodiesel-production-from-dende.html**

Threshing	Electricity	7.5 MJ	Carbon dioxide CO: (inorganic)	24.72kg
Digestion			Carbon dioxide COj (blotite)	801.37 kg
Pressing			Methane CH_4	0.04 kg
			Nitrous Oxide N 0?	1,-36 kg

Figura 33 Table 1. Life cycle inventory for the production of 1 tonne of palm oil during the cultivation phase.

	Entry		Output	
Processes		Quantity		Quantity
Seedling production	Seeds	40 units	Fresh fruit bunches	5000kg
Soil preparation	Fertilisers	82.15kg	Carbon dioxide CO_2 (inorganic)	9.01kg
Planting	Pesticides	0.69kg	Carbon Dioxide COJbiotic)	595.15 kg
Planting care	Diesel	0.84 kg	Methane CH^	0.02kg
Harvesting	Polythiophene	0.37kg	Polythene	0.37kg
	Dioxide Carbon CO_2	1647.5 kg		
	Water	2.23kg		

Figura 34 Table 2. Life cycle inventory for the production of 1 tonne of palm oil during the oil extraction phase.

Characterisation of impacts

The following table shows aspects of the oil palm production system

Problems	Analysis	Proposals	Action/Years Short: 1 Medium: 2 Long: 5
Monoculture	Lost biodiversity, rural exodus, unemployment, degradation of natural resources, lack of food, contamination of soil and water by pesticides.	Presenting options agroforestry and the advantages of consortium vegetables. Obtain financing through Pronaf	1 year
Land Statute Legislation Decree No. 59.566 of 14 November 1966. Use this law to make this	The lack of renewal in Brazilian legislation on land use, in a way that ends up benefiting	Mixed lease with profit share in cash or plant. Plus the intercropping of oil palm trees, which would only	1 year

new mo del viable lease	them. large investors, such as agro-industries.	benefit the legal owner of the land, means that the owner has several options for guaranteeing a livelihood for his family: 1- Rental value; Leasehold ; (Leases for oil palm cultivation and uses the spaces between the rows to cultivate	
		othercult ures p ermanent temporary). 2- 0 Surplus harvested from the owner's plantation can be commercialised locally or externally, encouraging growth in the local economy. 3- Labour: Small farmers can also be paid to look after the oil palm crop or join a small farmers' cooperative.	
Rural exodus	Due to a lack of means of subsistence and low wages, small farmers aban don their farms. land and open up space for agro-industry, whose sole aim is monoculture and economic productivity.	With options for their subsistence through renting and hiring labour and intercropping food crops for family sustenance, they guarantee the settlement of traditional communities and small farmers.	2 years
Few pro	Few centres of treatment what	To per	5 years

cessing centres vegetable. As oil palm has a short shelf life after harvesting, the process of extracting the products and by-products has to be carried out in order to ensure that they are safe.	overloads production by-products unproductivity.	fect production to set up small oil palm processing evenly distributed in partnership with the major oil palm industries.	
process quickly		It would guarantee the quality of products and increase local employment.	
Biomass	Excess by-products obtained in the production process. Waste of organic material.	Use of inert fibres for energy production. This could be financed and supported by large agro-industries. Completing the total utilisation.	1 year

Figure 35 Main aspects of oil palm production

Use of agrochemicals Conventional oil palm cultivation requires the use of agrochemicals, with the expansion of the crop tending to increase their use.	It would in crease contamination of the physical and biotic environments, je opardising the biodiversity, strengthening pests and creating a vicious circle.	Biological control and int ercropping repellent crops.	5 years
Conventional techniques	Soil degradation, water and air contamination, waste biodiversity.	Organic technique , initially high labour costs, but higher financial returns due to demand for organic	2 years

		products. Collective certification for direct sales to the few setting targets for obtaining comprehensive commercial certification.	
Lack of specific information and training for some crops and cultivation techniques.	Inadequate use of fertilisers, pesticides and herbicides, waste of area.	Increase the productivity and quality of the products grown. Using efficient, low-cost technologies in line with proposals and tests already carried out by experts in the field, through consultancy and free workshops.	1 year
High water utilisation.	A large amount of water is used in the palm oil extraction process.	As the region has a high rainfall rate, rainwater harvesting would be a way of minimising expenditure in the production sector. household utilities.	1 year

Figure 36 Table 21 - Representation and suggestions for improvement. Representation of **0**

product and by-product and their chain of impacts (Caires, Santos 2015).

CHAPTER 5

PROPOSALS

SAFs

Oil palm cultivation is a permanent crop, with production starting after 4 to 5 years and yielding up to 25 years. Because of these characteristics, there is a need to grow other vegetables, which is of great importance for keeping traditional families fed.

Researchers are studying the possibilities of intercropping and some have already been tested and have obtained satisfactory results, such as pineapple, manioc and beans, taking into account the characteristics of each plant as well as climatic and geomorphological features.

But according to other cultivars in common use in the region in the state of Pará, various alternatives can be highlighted, as long as the individual characteristics of each species are respected. It is also important to emphasise the region's economic activity in relation to the area available, as can be seen in the tables comparing cultivars taken from the IBGE.

In view of this survey, there are some ecologically sustainable suggestions such as the intercropping of species that have the power to repel certain pests and also the extraction of therapeutic and phytosanitary products such as Pimenta, Neem, Jaborandi and Andiroba, which can be intercropped to benefit the entire cultivation process.

The other options mentioned were for family sustenance and marketing, such as manioc, pineapple, papaya, cocoa, açaí, bananas and beans. Pineapple, manioc, bananas and chilli peppers stand out for their added economic value.

Workshops on oil palm cultivars, harvesting and production.

Palm oil is a vegetable that is rich in all aspects and is considered to be the largest oil producer in the world, beating soya.

But for this process to be improved, there is a need to train producers from traditional families to make the best use of this resource.

This takes the form of courses and workshops that can be sponsored by the agro-industries interested in the product, forging closer links of experience.

Local investment by agro-industries

Traditional families in the state of Pará are characterised by a low-income situation and need many basic services such as education, health and infrastructure such as electricity and sanitation. Through cooperative partnerships with agro-industries, it would be possible to programme these benefits effectively, bringing advantages to both parties involved and socio-economic and environmental development would be advancing.

Indications and improvements

- Rural medical care;
- Rural school from kindergarten to high school and EJA.

- Training workshops for women (entrepreneurship, traditional cookery, handicrafts with natural materials, the environment, organic gardening, recycling).
- Women's Cooperatives - entrepreneurs of locally produced products.

- Workshops on making biodigesters, composting and using organic waste for biomass energy, such as small thermoelectric plants.
- Piped water and sewage treatment. Prevention of diseases and epidemics

Mixed tenancy and intercropping

Lease the desired area by putting in the contract established rules such as intercropping, with the tenant growing the oil palm and the landowner growing the intercropped crop, in accordance with the Land Statute.

This guarantees the subsistence of small farmers and the security of biodiversity in an ecological way.

Intercropping model						
Vegetable	Type	Hectarees	Ton.	Value	Harvesting	Production
Pineapple	Temporary	10.777	320.478/un.	263.636	14/18 months	-
Cassava	Temporary	302.300	147	2.208.029	10/18 months	-
Beans	Temporary	43.941	30.737	65.493	60/120 days	-
Pepper	Permanent	13.958	30.885	365.334	90/120 days	-
Neem	Permanent	-	-	-	1 to 2 years	1/year
Banana	Permanent	43.510	585.943	411.926	14 to 18 months	all year round
Acai	Permanent	59,2	11.073	291.991	3 to 4 years	all year round
Dendê	Permanent	54.475	1.040.538	263.285	3 to 4 years	all year round

Figure 37 Caíres and Santos 2015 - IBGE data

Data on Neem was not found in IBGE data, but given the species' characteristics, even though it is exotic, it is adaptable to the Pará region, bringing benefits to agriculture, serving as a windbreak and organic matter and utilising phytosanitary active ingredients.

Gradual reduction in pesticides and fertilisers

As oil palm cultivation requires a lot of care due to pests, and requires fertilisation, set targets for reducing pesticides and replacing them with intercropping and intercropping with vegetable crops that have a repellent action. With intercropping, maintenance pruning should be carried out and the organic material used as organic fertiliser.

Some plants have the characteristic of fixing nitrogen in the soil, and you could also use these species for this nutrition.

Increased labour supply, improvements in health and education.

With local co-operatives comes an increase in jobs and consequently a need for medical care, which can be demanded by co-operative members. Co-operatives also bring learning about farming techniques and models, and these interactions between farmers and the co-operative only tend to grow and open doors to various sectors, which will be listed below:

Work for women: small vegetable gardens, handicrafts, traditional cookery, etc. Applying for funding to start and continue farming activities.

The possibility of partnerships with large agro-industries and local investment such as the initial oil palm processing centres. As the installation of a palm oil processing centre requires a large infrastructure, investment in sanitation, energy, job offers and improvements in health prevention would have a positive impact. This would already have a positive impact on the scenario, provided it was carried out with a good study and planning.

Partnerships with agro-industries

Setting up palm oil extraction plants would reduce transport to the mills and increase labour supply, productivity and product quality, as the extraction process requires a certain urgency after harvesting due to oxidation, as presented in an interview with EMBRAPA researchers. Utilising all the material obtained from the oil palm, including the waste that can be transformed into fuel for local electricity production and to keep the plants running. Much of the population lacks electricity, which would be another benefit for the population.

CHAPTER 6

Conclusion

With this work, it was possible to observe that development in the state of Pará is advancing at a rapid pace, but this does not indicate that this growth is distributed in a coherent manner. The need for public policies is extremely important if everyone's rights are to be guaranteed.

The advance of oil palm in the northern region of Brazil raises this question, and with it the need for these adjustments is eminent. Oil palm has all the advantages it needs to advance steadily, but for this to happen, integrating interests and thinking in the long term is of the utmost importance, because every positive impact has its negative impact and this needs to be carried out in a well-planned way so that problems don't occur that have already been seen with other types of crops that were initially promising.

Considering that oil palm is a perennial vegetable with great oil production power and produces by-products that are used in a wide range of products and that 100 per cent of all its waste is used, we can consider that oil palm is not very polluting and is a valuable product for the Brazilian economy, but we must not lose interest and investment in other means of subsistence and provision, such as food, in order to maintain food security and ensure environmental and social interests.

REFERENCES

BCB - Central Bank of Brazil 2015. Rural Credit 2015. Available at: https://www.bcb.gov.br/pt-br/#!/n/CREDRURAL . Accessed on: 2 May 2015 at 18:25.

BRAZIL. LAW No. 12.255, of 15 June 2010. Brazilian minimum wage for 2010. Available at: https://www.portalbrasil.net/salariominimo 2010.htm . Accessed on: 14 May 2015 at 8 p.m.

CHIES, Vivian. Dendê: Source of oil for biodiesel 2012. Available from Embrapa Agroenergia : http://sistemasdeproducao.cnptia.embrapa.br/FontesHTML/Pimenta/Pimenta c apsicum spp/clima.html accessed on 28/5 Accessed on: 3 May 2015 at 22:40.

CERQUEIRA , Wagner, F. Physical Aspects of Pará. Available at:

http://www.alunosonline.com.br/qeoqrafia/aspectos-fisicos-para.html . Accessed on: 7 Apr. 2015 at 22:17.

EMBRAPA - Brazilian Agricultural Research Corporation. **By CHIES, Vivian (News) 2015.** Scientists extract nanometric fibres from cellulose-base for new materials. Available at: https://www.embrapa.br/busca-de- noticias/-/noticia/1497582/nanofibres-from-cellulose-can-be-extracted-from-dende-residues . Accessed on 28 May 2015 at 16h.

FUNAI - National Indian Foundation. Types of Indigenous Lands.
Brasília DF 2015. Available at: http://www.funai.qov.br/index.php/indios-no- brasil/terras-indiqenas . Accessed on: 2 May 2015 at 11:27 am.

Government of the State of Pará. Available at: http://www.pa.qov.br. Accessed 12 Apr. 2015 at 17:57.

Government of the State of Pará. General Information on Ecological-Economic Zoning, available at : http://www.semas.pa.qov.br/servicos/zee/. Accessed 12 Apr. 2015 at 16:45.

IBGE - Brazilian Institute of Geography and Statistics.
Agricultural Census 2006. Rio de Janeiro: IBGE, 2006. Available at: http://qoo.ql/9VW51 K . Accessed on 20 April 2015 at 1pm.

Education Census 2012 . Rio de Janeiro: IBGE, 2012a. Available at:<http://www.ibqe.qov.br/estadosat/temas.php?siqla=pa&tema=educacao20 12> accessed on 24 April 2015 at 13h.

Crop Theme 2012 . Rio de Janeiro: IBGE, 2012a. Available at http://www.ibqe.qov.br/estadosat/temas.php?siqla=pa&tema=educacao2012 Accessed on 24 April 2015 at 13:30.

Tema Lavoura Permanente 2013.Rio de Janeiro: IBGE, 2013a. Available at : http://www.ibqe.qov.br/estadosat/temas.php?siqla=pa&tema=lavourapermanent e2013 ■ Accessed on: 24 Apr. 2015 at 15:18.

Temporary Crop Theme 2013.Rio de Janeiro: IBGE, 2013a. Available at: http://www.ibqe.qov.br/estadosat/temas.php?siqla=pa&tema=lavouratemporaria 2013 . Accessed on: 24 April 2015 at 15:00.

2010 Demographic Census Theme. Rio de Janeiro, 2010a. Available at: http://www.ibqe.qov.br/estadosat/temas.php?siqla=pa&tema=censodemoq2010 indiq univer .Accessed on: 25 April 2015.

Theme Plant extraction and forestry 2015. Available at: https://cidades.ibqe.qov.br/brasil/pa/pesquisa/16Z0 Accessed on 25 April 2015.

MMA - Ministry of the Environment 2015. Rural Environmental Registry. Available at: http://www.mma.qov.br/desenvolvimento-rural/cadastro- ambiental-rural. - Accessed on: 16 May 2015 at 1pm.

MAPA - Ministry of Agriculture, Livestock and Supply. The MAPA Chambers - A democratic and transparent instrument for dialogue with society Brasília DF 2009. Available at: http://www.aqricultura.qov.br/assuntos/camaras-setoriais-tematicas/arquivos-publicacoes-camaras-sectoriais/camaras sectoriais.pdf Accessed on: 16 May 2015 at 18:38.

MONTEIRO, K.F.G. et al 2015 - Oil palm cultivation as a production alternative for family farming and its inclusion in the biodiesel chain in the state of Pará : http://www.cfs.ifmt.edu.br/qet file/2000016/1000473/2/ . Accessed on: 03 May 2015 at 14:00.

SEMAS - State Secretariat for the Environment and Sustainability - Pará State Government 2015. Available at: https://www.semas.pa.qov.br/ Accessed on: 20 May 2015 at 1pm.

SOUZA, Jonas. Dendê (Extract from CEPLAC News - October/2000) Available at: http://www.ceplac.qov.br/radar/dende.htm. Accessed on: 05 Apr. 2015 at 15:57.

STCP - Engenharia de Projetos Ltda. Characterisation of Oilseeds for Biodiesel Production 2006. Available at: https://www.embrapa.br/busca-de- noticias/-/noticia/1487572/dende-fonte-de-oleo-para-biodiesel . Accessed on 5 April 2015 at 16:53.

Videos

ALVES, B. A. Field day on TV. Intercropping oil palm 2015 (42m26s). Available at: <https://www.youtube.com/watch7v- rPvYyPYL5A#action=share >. Accessed on 16 June 2015.

CONNECTION, sciences. Oil palm accumulates carbon and increases soil nutrients in agroforestry systems 2015. Available at: <https://www.youtube.com/watch?v=jOGvOQ3CkVq#action=share> Accessed on 16 June

2011.

CEPLAC - Artisanal palm oil production 2012 (7m44s). Available at: Accessed at: https://www.youtube.com/watch?v=qJ06EpT67nA Accessed on: 5 May 2015 at 17:50.

COCOA MARKET. Dendê in Valença - Field Day on TV 2012. Available at: https://www.youtube.com/watch?v=mW-tp5NR3xq Accessed on: 26 March 2012.

ROQUE, Araújo. Dendê 2011 (11m12s). Available at: <https://www.youtube.com/watch?v=Go2nwHFe01Y> . Accessed on 16 June 2011.

TV - Field Day. Production and use of palm biodiesel 2009 (6m07s) Available at: <https://www.youtube.com/watch?v=GzRIK39iffE> Accessed on: 16 June 2015.

TVNBR - Sustainable Oil Palm Production Programme 2011 (59m02s). Available at: <https://www.youtube.com/watch?v=Af8SZTaBHto> Accessed on: 16Jun.2015.

APPENDIX

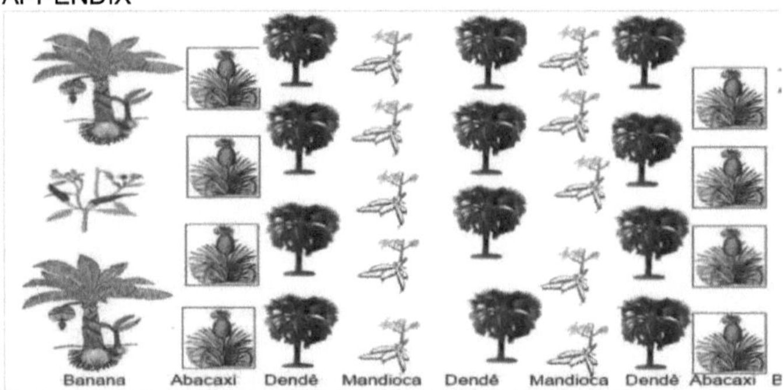

Figura 38 APPENDIX A - Intercropping models Source: Caires, Santos.2015

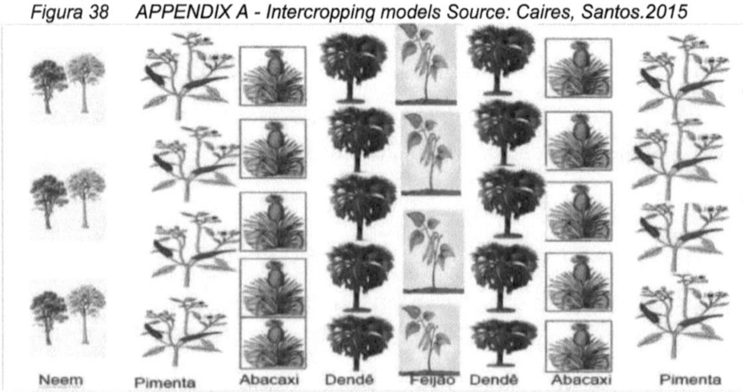

Figura 39 APPENDIX B - Intercropping models Source: Caires, Santos. 2015

Figura 40 APPENDIX C - Intercropping models Source: Caires, Santos.2015

MIX
Papier aus verantwortungsvollen Quellen
Paper from responsible sources
FSC® C105338

Printed by Books on Demand GmbH, Norderstedt / Germany